Build Your Own PC

Mark Dufour

authorHOUSE®

AuthorHouse™
1663 Liberty Drive, Suite 200
Bloomington, IN 47403
www.authorhouse.com
Phone: 1-800-839-8640

First published by AuthorHouse 11/19/2008

ISBN: 978-1-4389-2589-9 (sc)

Printed in the United States of America
Bloomington, Indiana

This book is printed on acid-free paper.

ACKNOWLEDGEMENTS

I would like to thank my sister Diane for helping me with the "about the book" page that I had to write. I would also like to thank my Nephew Erik, and Diane, for proofreading the book.

INDEX

DISCLAIMER

Since you will be working with 120 volts, and low voltages, You will need to use CAUTION. Caution not only from voltages, but from static electricity. Since static electricity can damage PC components, you MUST use an anti-static wrist strap when handling PC components. If you fry a part, you are on your own ! You must follow the instructions in this book to the letter. It is also advisable to read on a bit when doing the procedures in this book. That way you know what's coming up next, or cautions for what you are doing. The author, publisher, representatives of the author, or representatives of the publisher assume no liability for fried parts, damage to your PC, someone else's PC you are working on, or electrocution from plugging things into outlets, old or outdated wiring, errors in house wiring, or bad grounds, etc... If you feel uncomfortable doing the things in this book, then maybe you should buy a PC, or hire someone to build one for you. The author, publisher, representatives of the author, or representatives of the publisher also assume no liability for lost data, misrepresentation of persons, companies, corporations, or omissions.

INTRODUCTION

Since the introduction of the first PC, things have changed a lot. But the fact is, it is easier now than ever to build your own PC. Lots of hardware compatibility issues have been ironed out. USB now works pretty solid, and graphics formats have been intergrated into the PCI x16 V2 express bus. When I built my first PC in 1999, I went through 7 monitors, 4 CD drives, and 2 sets of speakers before I had a working PC. Everything seemed to conflict with everything ! Now it's a lot different. Hardware actually works 95% of the time when you install it, At least for windows XP®. I tried Windows® XP® 64 bit, and had some trouble installing some programs. Most utilities simply didn't work. I tried Vista® about 6 months ago, and I was not impressed. Vista® drivers were scarce especially for older hardware. I returned it. It was one of those free trial offers. If you were to ask me what Operating System you should go with ? I would have to say at this time, stay with Windows®32 bit XP®.

If you build your PC yourself, it would be modular. If you decide to later add something, or upgrade some part of it you can, plus you get the satisfaction of building it yourself while saving money. You can't do that with all pre-assembled PC's that you buy today. Just trying to open the case on some of these new PC's can be very challenging.

Chapter 01

CONSIDERATIONS

How much is this going to cost me ? Well if you don't already have them, you'll need the basics like: Monitor, keyboard, mouse or trek ball, printer, and speakers. You will need to buy the case, fans, (if not included with the case), power supply, (if not included with the case), DVD, and hard disc drives. Then figure how much horsepower do you need, and go from there. You will need to buy the CPU (processor), CPU fan, (if not included), motherboard, and memory. You want to buy as much processor as you can afford. Do you want 2 video cards, one, or none (if you have onboard video)? You would use 2 video cards in tandem (SLI) for extreme 3D gaming, or a 2 video card 3 monitor setup for a flight simulator. Get the latest video card model that you can afford. I wouldn't go over board and spend more than $275 though. New video cards come out roughly every 4-6 months. If you plan on playing the latest and future video games, you probably want to go with a video card, and not onboard video. That way you can upgrade that too. Last, a sound card if you decide to go with that option, and not onboard sound. If you want high quality 5.1, or 7.1 surround sound (includes side speakers), or you plan on using MIDI, then I suggest you go with an audio card, and not onboard sound. If you're going to use your PC for creating music, then get a Sound card that can record and playback at the same time. It's called a duplex sound card.

FUTURE CONSIDERATIONS / OBSOLESCENCE

They say your PC is obsolete when you take it home. If you buy one, then the video and sound are probably built onto the motherboard, and you can't upgrade it unless you have empty PCI slots. If not then you have to change the motherboard. The power supply is probably too small to handle anything bigger anyway. You will run into problems like this if you try to upgrade. If you build your own, you can upgrade any part of it later. Will my new PC that I build be cheaper than one that I buy ? It depends on how much horse-power that you want. It also depends on how many components that you have already. If you already have a keyboard, mouse, speakers, and a monitor, and you are happy with those, you can use them again on your new PC. If you decide to build a top end 3D gaming machine, then yes you can save money ! If you just want something to surf the internet with then, it would probably be cheaper to buy one. It really depends on if you want to play the latest, and future PC games, or just want to build a PC !

Chapter 02

PRICING EXAMPLE

Let's say you want to build a top notch gaming machine, and already have a monitor, mouse, keyboard, speakers, printer, and scanner. This leaves you with buying:

ATX Case- the more expensive ones already have fans installed. Typical -------- $100
Power supply - sometimes the case will include the power supply------------------$130
2 Gigabytes of Corsair® memory -- $ 80
3.2 Ghz AMD® Athlon 6400+ DUAL CORE Processor (with cooling fan) ---- $119
Asus ATX Motherboard --- $170
Western Digital® 500GB Hard Drive --- $ 90
Soundblaster® X-FI® Extreme Gamer® Sound card ------------------------------- $ 90
(Not necessary if you have sound on the motherboard)
EVGA® geforce® 9800GTX+Video card -- $200
(Not necessary if you have video on the motherboard)
Sony® DVD drive ---$ 70

	SUB TOTAL	$1049
If you don't need the video card then subtract		- $200
	SUB TOTAL	$849
If you don't need the audio card then subtract		- $ 90
	TOTAL	$759

NOTE: If you have onboard audio, and video, your motherboard will cost you about $40 more.

NOTE: You should also buy an outlet strip with line spike protection, and a ground LED. The ground LED tells you if you have a proper ground. Buy an outlet strip that has a switch.

NOTE: The power supply that you get free with a case is not usually as good of a quality as a power supply bought separately. The quality that I'm referring to is the filtering, and wattage rating.

Chapter 03

CHOOSING COMPONENTS

Processor

Buy as much processor as you can afford, and get a dual core. Look for at least a 3.2 GHZ Processor or CPU. Modern games require 3Ghz, and are exceeding this requirement. I've used AMD® in every one of my systems. And have never had a Problem with any one of them. It's a matter of preference, and price ! What socket type Should I buy ? I've had good luck with AM2. (AMD 64bit, dual core). I'm presently using an AMD ATHLON® 6400+ dual core 3.2 Ghz Processor for my system. If you want to go quad core then you would need socket AMD AM2+. Choose the processor first for your system.

Motherboard

64 Bit vs. 32 Bit. First it must be approved by the processor Manufacturer. You can look this up at the processor manufacturer's website. Second if your building a killer game machine you might consider getting a 64 bit motherboard. Motherboards come in three sizes, ATX, Micro ATX, and SMALL. The SMALL, and micro ATX can't accommodate all the possible features of ATX simply because of size constraints. If your all thumbs, go with ATX, as this will give you more room in the case to work in. I've had good luck with Asus®. I have an Asus® Crosshair® ATX 64 bit motherboard with a built in troubleshooting LCD that reads out the steps it is taking to power up. This is a very handy tool for first time builders. The crosshair is already two years old, but I still use it for my gaming rig. I also have an Asus® M3N72-D (32bit) ATX motherboard that I bought this year. You want to pick a Motherboard carefully. This ultimately determines what features you will have as far as hardware goes. In other words you have to pick your features first, and find a motherboard that will support your features. Actually it is the chipset that will determine what you have for features. You will want features like:

SATA II (This is serial hard drive support Version 2).
At least 4 Gigabyte memory support, and DDR2, or DDR3 memory support.
Dual channel memory support.
Dual PCI X16 express, Version2, with SLI support. (This is only needed if you want to run 2 video cards.)
64 bit Processor support. (Optional)
IEEE 1394, or 1394B support.
S/PDIF support. (This is for home theatre surround sound)
USB 2.0 (This lets you connect version 2.0 USB hardware.)
 Temperature/voltage monitor support

DON'T FORGET TO MATCH YOUR MOTHERBOARD SOCKET TYPE TO YOUR PROCESSOR SOCKET TYPE. YOU ALSO NEED TO MATCH YOUR MOTHERBOARD FORM FACTOR TO YOUR CASE FORM FACTOR.

Processor socket type for example would be AMD AM2®, and your form factor is the motherboard size. An example would be ATX or micro ATX. Choose your motherboard second for your system. IF you are confused, that's okay, buying a motherboard is a confusing, tedious process. One question arises ? What if I find a good motherboard, with everything I want, and I want a sound card, BUT it comes with onboard sound, do the two conflict ? YES, you have to disable onboard sound in the bios. Follow the instructions, and this will be taken care of. If you want to use a video card, then you have to disable the onboard video as well, (if you have onboard video on your motherboard).

Power Supply

For a general rule of thumb, get 500 Watts for a run of the mill desktop, and at least 750 watts for a gaming machine. If you're running 2 video cards, use a 750 watt, then be sure to check the power rail rating. Make sure it is approved by the Motherboard manufacturer. Check their websites.
NOTE: MAKE SURE THE POWER SUPPLY HAS AN ON/OFF SWITCH.

Power rail considerations. Make sure a 750 watt power supply has at least the following ratings.

+3.3volt	30	amps
+5volt	30	amps
+12volt	60	amps
-12volt	.8	amp
+5volt sb	2	amps (sb means stand by)

Choose the power supply third for your system. Choosing a power supply can also be frustrating. If you find that this is the case then just buy a Corsair 750HX, 750 watt power supply. This is what I have in my gaming PC. You can also visit Corsair.com to choose a lower wattage 500 watt power supply for a desktop. Corsair has a 5 year warranty on the 750 HX series.

Case

One of the most important parts of your PC is the case. This is what everything connects to ! You will want to buy a case that has USB and 1394 ports in the front. This way you can plug in USB devices from the front including, those neat memory sticks, or your digital camera. Get at least 2 optical drive bays, and 2 hard disc drive bays. The better cases have air filters that you can remove and clean. I prefer Antec®. They make cases that make working on things much easier, especially if you get an ATX size case. If you want a smaller case go with Micro ATX. Your motherboard also has to be Micro ATX. The more you over heat your processor the shorter it will last. So you want to pick a case with good airflow. The air should come in the front, and flow freely around the case, and exhaust out the back, or top, or both. Choose this component fourth for your system, and the rest of the components really don't matter in what order you choose them.

Memory
Buy at least 2-4 GIGABYTE of a good brand name like Corsair®.

Memory types - PC2 means DDR2   (DDR stands for double data rate)
                     PC3 means DDR3

Examples: PC2 6400 = DDR2 800MHZ (800 Megahertz)
                PC3 8500 = DDR3 1066MHZ

Match the memory speed for the maximum speed your motherboard will handle. You also want to buy memory in matched pairs, if you're getting two sticks. This will ensure a stable system. If you decide to add more memory later, get the exact same kind.

Hard Drives
Get two (optional) at least 500 Gigabytes, one for backing up. I prefer Western Digital®. Get at least 7200 RPM. Get the internal version(s). The reason I prefer Western digital is because of the quality, and the drives come with an erasing utility.

Optical Drives
DVD - You're going to want dual layer recording. I prefer Sony®. Get the internal version. Don't forget to match the color to the case color.

Sound
The standard for audio and MIDI is Creative Soundblaster®. I have a Soundblaster® X-FI Extreme Music. Why buy anything else that has to emulate the Soundblaster® standard ? If you skimp here, you might run into trouble down the road, like I did.

Video
The standard format for video cards today is PCI Express X16 Version 2.Get at least 512 Megabytes of video ram. Get one that supports Direct X® 10, and OPEN GL® version 2.1

Note: Direct X 10 is required by Windows® VISTA®

NVIDIA® vs. ATI® Here we go again. I've always used Nvidia®, and never had any problems. This is another matter of preference. Some motherboards have video built in. If the built in (onboard) video is suitable for your needs then by all means use it. If not then add a video card to suit your needs. You can always add a twin video card later, and run them in SLI mode. This will almost double your video speed. That is if you picked out a motherboard with dual video card support. I have two EVGA NVIDIA GEFORCE 9600GT video cards not running in SLI, because I need three monitors for my flight simulator. The video cards have 2 monitor ports each.

Fans

Quietness is a big factor here. Go with 120mm fans for the case. My case came with ANTEC fans. Don't forget the processor cooling fan. I have a THERMALTAKE® Get one that's approved by the Processor manufacturer. Get fans with the lowest DB(Decibel) value, as they will be more quiet. NOTE: The processor sometimes comes with the fan.

MONITOR LCD vs. TUBE

I prefer a tube monitor in all viewing areas with the exception of weight, and on time. A tube monitor is visible from almost any angle. I think this is important, especially if you're trying to show someone something on the screen, or standing up. A tube monitor is easier on the eyes, and has better resolution. When it comes to fine lines, the LCD monitor may not show them, but your printer will if it is a good quality printer !

Tube Type

Shadow mask® vs. aperture grille. I think the shadow mask® type tube is sharper, and easier to look at. Go with =<.26mm dot pitch. I have a Philips® model 201B4 with a 21 inch screen. I thought I would try a 19 inch LCD monitor but, I prefer my tube any day. I cannot recommend the Philips model that I have because of the fragile power switch design. I'm easy on stuff, and it broke in 2 years ! However the picture is very good. Luckily I'm an electronic tech and was able to fix the switch.

Keyboard

Curved vs. Straight layout – Once you get used to a curved one like Microsoft natural®, you don't go back. A curved keyboard feels quite natural. It took a while to get used to it though.

Trek balls

Reasons to get a trek ball: First they are faster for gaming. Second If you have shoulder problems, you don't have to move your shoulder around. They stay put ! I have a Logitech marble mouse® trek ball. They have 4 programmable buttons. I love this thing ! It comes with software that allows you to close programs with a single click when your pointer is anywhere on the program box. You can also shut down your PC this way.

Mouse

If you decide not to go with a trek ball, make sure you get a thumb wheel, and three or four buttons.

Printers

I've only owned Epson Stylus Photo® series inkjets, and I can vouch for them. Look for at least 1440 by 5760 dots per inch.

Scanners

Here again, I've only owned an Epson Expression® 1600 series. Look for at least 800x1600 resolution if you plan on doing large photos like 8.5 x 11's.

All in ones

I can't recommend a brand here as I don't own one.

Speakers

I've had good luck with Yamaha®, and Altec Lansing®. Here again the power switch broke after about 3 years on the Altec Lansing®'s. Luckily, I was able to fix that too ! If you're looking for surround sound, Logitech® model X-540 is a good surround setup for $79. That was the price a year ago. It also sounds great for only 70 watts rms of power. Multiply watts (r m s) by 1.414 to get peak watts.

Chapter 04

Mobo Combos

Another way to get a better deal buying parts is to buy your motherboard, processor, and memory together. This way you can get the combo pretested. This is a big plus for first time builders. This reduces the chance of getting a DOA mobo (dead on arrival).

Bare bones

Here is yet another way to save more money. A bare bones setup includes the case, power supply, and motherboard. Most of these motherboards have onboard audio, and onboard video.

YES, PC parts also come with rebates !

Chapter 5

GETTING STARTED

Tools

You will need a medium philips screwdriver, and a small pair of needle nose Pliers. That's it !

DO NOT OPEN ANY BAGS, OR BOXES UNTIL YOU ARE STRAPPED IN AND READY TO INSTALL THE COMPONENT, with the exception of the power supply.

INSTALLING THE POWER SUPPLY

Install the power supply into the case. Do this buy screwing it into place so the switch is facing out toward the back wall of the PC case. Align the holes. Use the 4 wide head philips head screws provided, and screw it in from the outside of the case. Some cases have an inside pedestal to mount the power supply on, so you will have to put the screws inside the case. Some cases come with manuals, read that too. The power supply can only go in one way. See figure # 1, Power Supply Location.

CAUTION: THE PC MUST BE PLUGGED INTO A GROUNDED OUTLET, OR GROUNDED OUTLET STRIP TO ESTABLISH GROUND. THE PC POWER SUPPLY SWITCH, AND OUTLET STRIP SWITCH SHOULD BE TURNED OFF WHEN YOU ARE WORKING ON YOUR PC !

STRAPPING IN

You will want to buy an anti-static wrist strap to work with the PC components. Computer components are static sensitive, and you can fry them just by picking them up ! You can buy these online, or @ Radio Shack®. Put the wrist strap on either wrist, and connect the alligator Clip to any bare metal on the PC case for your wrist strap ground. You must plug in the power supply to have a ground. (Make sure the switch is off !) It's best if you have a grounded surge suppressing outlet strip with a power on light on your table. Buy the kind with a switch. That way you can turn the power off to the whole PC, and monitor when you want to. This will also help save electricity.

NOTE: Even though the power switches are off, you are still grounded if you are plugged in. That is if you are using a power strip with a grounded plug, and you have modern wiring.

NOTE: You may want to jot down your component info on the sheet provided in the back of the book for reference. (part #'s log) Do this BEFORE you install every component. This comes in handy for rebates, or warranties. You may also want to write down the date purchased.

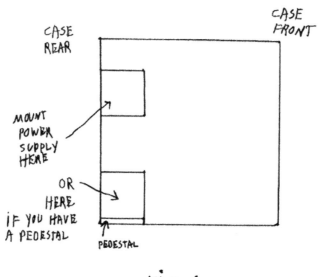

CASE
REAR

CASE
FRONT

MOUNT
POWER
SUPPLY
HERE

OR
HERE
iF YOU HAVE
A PEDESTAL

PEDESTAL

FIG 1

Chapter 06

PUTTING IT ALL TOGETHER

NOTE: You will need a clean, dry, flat surface. It is recommended that you do not use a surface that could possibly contain steel wool, bits of metal, or solder flakes, as this could be deadly to the motherboard. First you will need to strap in and ground yourself. (See STRAPPING IN chapter # 5), and unbox your motherboard, processor, memory, and video card (if you have one). Leave them in their bags, or anti-static containers for now. Now would be a good time to look over your motherboard book and familiarize yourself with the layout of your motherboard. You will need to know which socket your type of video card plugs into (if you have a video card). For example PCI EXPRESS, or PCI EXPRESS #1 (if you have dual video card capability).You will also need to know where to install the processor, and memory. Now unbag the motherboard WITHOUT TOUCHING ANY OF THE COMPONENTS, AND HANDLE ALL CIRCUIT BOARDS BY THE EDGES ONLY. Orientate it with all the components facing up. Now orientate it so the edge of the motherboard that has all the sockets faces left.

INSTALL PROCESSOR
Look at the processor (CPU) socket on the motherboard, and look at the motherboard layout in the motherboard book. There is a little metal bar mounted on the side of it. Lift the bar straight up. You may have to nudge it away from the socket direction slightly. Now notice that the holes in the socket have a few missing holes on one corner. Take the processor out of it's package, and handle it by the edges only. Be careful not to touch or bend the pins. Align the CPU so that the corner that has a few missing pins aligns with the motherboard's missing holes corner. Simply place the CPU in the holes. It should drop in. Now take the little metal bar and put it back in it's original down position.

INSTALL HEAT SINK
This is probably the hardest part to describe, AND DO in all this book ! Apply an even amount of heat sink compound onto the flat center of the CPU metal cover. Apply ~ .5 millimeter thickness of compound. See figure 2 processor heat sink placement. The fan/heatsink combo also has a lever that you have to move in the upright position. You also have a metal H shaped metal strap that has square holes in the sides of it. At this point, "motherboard" and the word "mobo" are used one in the same. Look at the plastic piece around the CPU on the mobo. It has 4 plastic notches that stick out. Align the metal strap so that the square holes align with the plastic notches on the mobo. DO NOT LET THE HEATSINK COME INTO CONTACT WITH THE PASTE YET. Now Place the

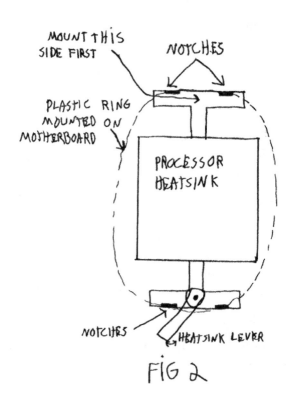

MOUNT THIS SIDE FIRST

NOTCHES

PLASTIC RING MOUNTED ON MOTHERBOARD

PROCESSOR HEATSINK

NOTCHES

HEATSINK LEVER

FIG 2

heat sink squarely onto the CPU, and hold it there with one hand. Slide the metal bracket so the end without the lever goes over to the plastic notches. With the other hand (you may switch hands) put the metal strap onto the plastic notches by slightly bending down on the metal strap. Now align and place the other end of the metal strap with holes onto the notches. Once in place, you must push down on the lever until it is in place. You may have to fiddle with it to get the metal strap down onto the notches. It is okay to switch hands if you DO NOT let go of the heat sink. Try not to move the heat sink around on the processor, as you will squeeze out the heat sink compound. Connect the fan connector to the CPU connector on the mobo. See your mobo book for the location of this connector, as all motherboards are different.

INSTALL MEMORY

See Figure 3. Read your motherboard book memory installation procedure to determine which socket(s) the memory will go into. Unbag the memory, and locate the memory sockets on the motherboard using the motherboard book. On each end of the memory sockets there is a tab. Push the little tabs down towards the motherboard. Install the memory into the memory socket(s) deep down enough so that the little tabs start to rise up into the memory sticks hole on the edge of the memory stick. The memory will only go in one way, you will have to orientate the memory and align the memory stick bottom notch to line up with the socket bottom notch. Then bring the little tabs up on each end until they go into place. If the tabs don't go into place (tab tooth inserts into the notch on the side of the memory stick) then the memory stick isn't seated deep enough. **NOTE: Do not press too hard on the Mobo, as it may CRACK !**

PREPPING THE CASE

Lay the case down on the right side. Keep the top of the case at the top. Open the left side, and remove the left cover, (if you are allowed to remove the cover).

STANDOFFS

If the case does not have standoffs for the motherboard to sit on installed, (now on the bottom of the case), then you will have to install them. In the bag of screws, use the brass, spacers. For an ATX mobo use nine standoffs. If your mobo is micro-ATX then mount six standoffs. Use the holes on the right side panel (now the bottom), of your case to mount the standoffs. You will mount the six standoffs for the micro-ATX in the holes closest to the back of the case. For ATX use all nine holes for the standoffs. The standoffs provide a space between the case right side panel (now the bottom), and the back of the mobo. This way the mobo doesn't short out.

MOUNTING THE MOTHERBOARD

Mount the backplane. This is the metal shield that has all the holes for the mobo external connectors. This snaps into place on the back of the case. Now mount the mobo. Align the external connectors on the left hand side of the mobo to the backplane.

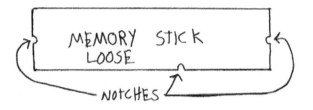

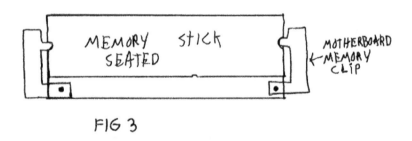

FIG 3

Sometimes, there are two metal tabs sticking out of the backplane that you have to be aware of. Some of the mobo external connectors fit between these tabs. Make sure the mobo does not move around on the standoffs, as this may damage the mobo. Let the mobo enter the backplane first in between the metal tabs, then let the mobo rest on the standoffs. You will have to press down on the mobo to keep the mobo in place. Screw in the mobo on the standoffs. Do not over tighten the mobo screws. Use 6 screws if the mobo is micro ATX, and 9 screws if the mobo is ATX size.

INSTALL MOTHERBOARD POWER CABLES

Connect the power supply plug with the most wires to the matching motherboard socket. It will only go in one way. Next connect the 4 or 8 wire motherboard power connector from the power supply to the motherboard. Look at your mobo layout page in your mobo book to see if it requires a 4, or 8 pin power connector. This is the ATX +12 volt motherboard EPS power connector. It can only go in one way regardless if it is 4 or 8 pin. Connect any temperature probes that you may have to the mobo. See your mobo book for the location of these temperature probe connectors.

Case connections

Refer to your mobo book "system panel connections" This is a 20 pin connector. If your all thumbs, this is the time to use your needle nose pliers. Connect the power switch, reset switch, hard disc led, speaker (if you have one), and system power led. The white wire is the negative. This means that all the white wires with no stripes must connect to negative. The speaker may be black and red, in this case connect the black to negative.

Fan connections

Connect all the fans to ONE of the 4 wire power supply connectors. See figure 6. If you have 3 wire fans, then the fans will connect to the motherboard. See your mobo book for fan connections, and the location of those connectors.

INSTALL VIDEO CARD - (If required)

Unbag the video card. DO NOT TOUCH THE GOLD FINGERS. Plug it into the motherboard. The video card can only go in one way. The video card will be perpendicular to the motherboard. The edge of the video card with the steel tab will be facing left. The steel tab has a smooth, flat tooth side, and a mounting side. Align the tooth side to go into the slot in the case. The video card edge with the little gold fingers will be facing down, and will plug into the motherboard PCI EXPRESS socket, or PCI EXPRESS #1(if you have dual PCI EXPRESS). Screw in the video card with the screw provided (it will be one of the larger ones). Plug your monitor into the socket on the edge of the video card with the steel tab. If there are 2 identical sockets, use the bottom one. Plug the 6 wire power connector to the edge connector opposite the steel tab on the other end of the video card. If you plan on adding two video cards, add one card for now.

Replace or close the left cover, and stand the pc up. It should still be plugged in, and turned off at this point.

Chapter 7

BOOT SEQUENCE

Connect your keyboard to the socket with the keyboard symbol in the back of your case. If you have a USB keyboard, use the rectangular to round plug adapter that came with it. Connect your mouse the same way using the rectangular adapter that came with the mouse. Connect the mouse to the mouse socket in the back of the case. Turn on the power strip switch, and the Monitor power switch.

Turn on the power supply switch, and your PC power switch. Your new PC should come to life with a brief information screen, and a little white bar should next appear at the upper left. Your monitor SHOULD THEN DISPLAY "BOOT ERROR INSERT SYSTEM DISC". This is NORMAL because you do not have a hard drive connected yet. If you got a boot error, you can pat yourself on the back, GOOD JOB ! If you did not get a boot error then go to CHAPTER # 8 <u>TROUBLESHOOTING MOTHERBOARD, PROCESSOR, MEMORY, AND VIDEO SECTION.</u>

Bios adjustments

NOTE: A word about BIOS. Don't fool around here in the BIOS and press keys that you're not supposed to. You may lock up your PC, and need to have it serviced.

Now we need to go into the bios and check to make sure the CD/DVD drive will boot before the hard drive, so we can install Windows®. Press and let go the reset switch on the front of the PC, while holding the delete key. The bios screen should come up. Go to the boot section, and make sure the CD/DVD drive boots before the hard drive. If you plan on using a floppy drive, this will need to boot before the CD/DVD drive. Consult your mobo book.

Chapter 8

TROUBLESHOOTING MOTHER BOARD, PROCESSOR, MEMORY, AND VIDEO

NOTE: most motherboards have some type of power indicator, or LED that shows the motherboard is on. If you do not see an LED on your motherboard, then don't worry about it.

Does an LED light up on the motherboard ? If not, you do not have power going to the Motherboard. Did you remember to connect the motherboard power connectors ? Is the power supply switch on ? If yes, then check your power strip for power. If you believe you have power going to your PC, then re-read your mobo book for memory insertion order, and double check everything (including front panel connections). Turn off the PC power switch, and the power supply power switch. You may even use only one memory stick for now just to test it, (if you have a set). If it doesn't work, then use the other memory stick only, and try the boot sequence again. If your mobo LEDS light up, and you do not get a boot error, then one of three things is faulty: A: mobo   B: memory   C: CPU   in which case you will have to return the MOBO, or MOBO combo. Many steps have been taken in this book to make sure you don't end up with a dead PC. However, you may end up with what's called a DOA motherboard. I personally have had this happen to me. Jot down the serial #, and return it return receipt requested. DON'T forget to remove the heat sink, the processor, and memory if you bought them separately.

NOTE: TO REMOVE THE VIDEO CARD SEE FIGURE # 4. There is a plastic tab that you must hold out of the way to remove the card, on most motherboards.

You may also try the memory in another pc to see if it works, but it must be the same pin count, and DDR type. The new memory must be equal to or slower in speed than the maximum of the PC you are trying it out in.

NOTE: You may also use another PC to check your processor.

NOTE: The PC that you use to try the processor out in must be able to accommodate the speed of the new processor, same socket type, etc...

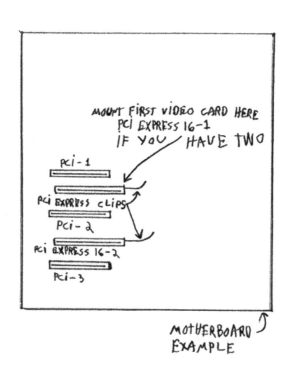

FIG 4

Chapter 9

TEMPERATURE / VOLTAGE TEST – VIDEO/AUDIO DISABLE

While still in the Bios screen. If you bought a video card, and your motherboard has onboard video, (in the main menu) use the arrow keys to go over to the "advanced" menu, or something like that. There should be a menu under it that says, "onboard device configuration", or something like that. All Mobos, and mobo books are different. Go down to onboard video, press enter, and follow the instructions on the screen to disable the onboard video. This usually is just an enter and, up or down arrow button press. Do the same thing if you bought an audio card. You have to disable that as well.

Press escape until you see the main menu, and then right arrow keys until the cursor goes over to the menu that says either hardware monitor, temperature monitor, or temperature / voltage, power, or something like that. (Motherboards will vary, read your motherboard book.) Go down to CPU temperature. (Move the down arrow key to it, and press enter if it does not display the temperature). The temperature should be 100 to 130 degrees F, the lower the better. Wait a few moments if it is below this and climbing. If it reaches a much higher temperature, (like 135 or higher), shut the PC down, turn off everything, and re-apply the heat sink compound a little thicker. It probably got squeezed out while mounting the heat sink. If it stabilizes at an equal to or lower temperature than 130 degrees F, you're good to go. My desktop PC runs at 104 degrees F(~40 degrees C). My gaming rig runs somewhat warmer at 122 degrees F, (~50 degrees C). The reason for this is the higher wattage processor.

Now look at the voltage monitor. All the voltages should be somewhat close to the stated voltages. You will have to send back the motherboard if you see any voltages in red. Press escape, until it says "exit" saving changes. Then save and exit. Shut everything off.

NOTE: If you have to send the motherboard back, don't forget to remove the heat sink, the processor, and memory, if you bought them separately.

Chapter 10

FINISHING THE HARDWARE

MOUNTING THE HARD DISC

Turn off your PC by pressing the PC power switch. Turn off the power supply switch. Turn off the outlet strip switch. Mount your hard drive(s) into the case. These screws are usually smaller in size. You might have to connect rails to it, then slide it in. See figure # 5. Refer to your mobo book for determining which SATA connector is #1. Connect your first (top) hard disc drive SATA connector to SATA socket # 1 on your motherboard Using the SATA cable. Your second hard disc drive will connect to SATA #2. (If you have a second drive). Plug in the power connector for each drive. See figure # 6

NOTE: If you have two hard drives, and an older version of XP® (lower than SP2), you will have to connect only one hard drive for now. After XP® is installed, then power down, shut everything off, and install the second hard drive. The reason for this is the drive ordering sequence of XP®. You will end up having your CD/DVD drive lettered E: You want your CD/DVD drive lettered D:

DVD

Mount the DVD drive into it's bay using the four screws provided. These screws usually are the smaller size. If your case requires mounting rails for the DVD drive, then see figure # 5. See figure # 6 for power connection. Connect the wider flat ribbon cable to the DVD drive. It can only go in one way. Use the end connector at the shorter end of the cable. Connect the other end of the flat ribbon cable (longer end) to the mobo socket labeled EIDE or IDE. Pay attention to pin #1 on the mobo EIDE connector, and Pin #1 on the DVD cable, the cable must be orientated correctly. The ribbon cable usually has a red or black stripe indicating the pin #1 side. Do NOT use the middle connector on the ribbon cable, this will connect to the second CD/DVD drive, if you have one.

NOTE: Do not confuse this mobo EIDE socket with the mobo floppy disc socket. They look the same, but are different lengths. The mobo floppy disc socket is usually marked floppy.

FLOPPY

Mount the floppy drive the same way as the hard drives. These screws are smaller also. You might require smaller rails if your case requires them. See figure # 5. Connect the floppy power connector. See figure # 7. Now connect the less wide flat ribbon cable (short end), to the floppy drive connector. Now connect the other end of the floppy flat ribbon cable (long end), to the mobo socket marked "floppy" paying attention to pin #1 on the mobo floppy connector. Pin #1 on the floppy must go to pin #1 on the

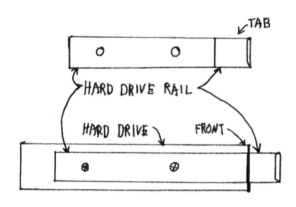

TAB

HARD DRIVE RAIL

HARD DRIVE   FRONT

FIGURE 5

FIGURE 6

EIDE HARD DRIVE CONNECTOR
DVD CONNECTOR
CD CONNECTOR

[0 0 0 0]

POWER CONNECTOR

SATA HARD DRIVE CONNECTOR

[▭▬▬▬▬]

POWER CONNECTOR

NOTE: IF USING SATA ON HARD DRIVE
DON'T USE THE EIDE CONNECTOR.

FLOPPY DRIVE
POWER
CONNECTOR

FIGURE 7

motherboard. The ribbon cable usually has a red or black stripe indicating the pin #1 side. The middle connector on the floppy ribbon cable is not used unless you have 2 floppy drives. You may have to buy this ribbon cable separately, if you want to use a floppy drive.

AUDIO CARD (If necessary)

Read the manual for your audio card to see if it requires a power connector.

NOTE: The audio card installs into a pci socket, NOT a pci EXPRESS socket. (Usually white). Read the motherboard book on PCI sockets. If the last one is not a slave then use that one. The reason for this is you will have more air circulation between the video card(s) and Audio card. If the last PCI socket is a slave then use the PCI socket one closer to the video card. If your audio card requires a power connector, then connect it using one of the power supply cables. (See figure # 6).

NOTE: At this point if you run out of cables from the power supply then you may use the cable that the fans are on. Simply leave the fans connected (doubled or tripled up), and plug this cable into the audio card.

NOTE: If you have an audio card, you must use the audio cable that connects the audio card to the DVD drive to hear the music if you play CD's on your PC. See your audio card book for connecting this cable. This cable should come with your audio card. If not then you will have to buy this cable separately. Refer to your audio card manufacturer.

Chapter 11

SOFTWARE INSTALLATION

Making a Driver disk

Before you begin to install windows, you should make a driver disk. You will need another PC to do this. Download from the internet, your latest drivers for your type of video and audio card. (If you have a video, or audio card). Do this also for LAN, BIOS, SATA driver, CHIPSET driver, TEMPERATURE PROBE software, PROCESSOR driver, PRINTER driver, SCANNER driver, or ALL- IN- ONE driver, and drivers for any other peripherals that you may have. Make a DVD with all of these drivers on it. Also include windows service packs. There are 3 of them. Do not put the drivers in folders. This way, you will have everything you need to update all of your hardware in one shot after installing windows. After installing windows is the best time to update your drivers because, as windows is installing hardware, it will ask you for the drivers it cannot find when you first start up after installing windows. As you download the drivers, when it asks you to save or open, click save. Rename the driver BEFORE you download it. Also, add the driver version along with the name. They will give you the version before you download it. For example name your video driver, "video_157.16".

Before you Begin

Boot your mobo again without holding down the delete key, and check the Fans to see if they are working, and check airflow direction. The air should blow in from the front, and out the rear or top, or both, out of the case. Leave the fans on high for now (if you have this option), until you see your system stabilize in temperature. This takes at least one half hour.

Windows® Basic Installation -

XP® 32 bit (with your PC still on)
Step 1 Insert the windows® installation CD, and keep the space bar pressed.

Note: If the system does not boot to the Windows® CD, then follow the steps for BOOT SEQUENCE Bios adjustments (chapter 7) otherwise continue to the next step.

Step 2 Windows® will load preliminary files.
Step 3 "Welcome to PC setup" - Press ENTER
Step 4 "License Agreement" - Press F8
Step 5 "Windows® Setup" - Press ENTER
Step 6 Press UP ARROW for quick format, Press ENTER

Step 7 Setup should start copying files, if it doesn't press ENTER. The PC should
         Restart by itself, and continue setup by itself.
Step 8 "Region and language options". Review and change, or click next.
Step 9 Enter name, or organization, and click next.
Step 10 Enter product key, and click next.
Step 11 Enter name, and, or click next.
Step 12 Enter Date and Time if not correct, and, or click next.
Step 13 "Network settings", click next. Windows® will restart. If you have a new
         version of Windows®, a display settings box will appear. If you can read the
         box, click okay. To continue, click next.
Step 14 Click on the white circle next to "turn on auto update", click next.
Step 15 "who will use this computer", Enter names, click next. If no one else, put your
         Name, click next.
Step 16 Click finish.

ADDING SERVICE PACKS / DRIVERS

Insert the software disc that you made, and install any service packs that are needed. A new copy of windows XP® usually has service pack 2 included, and maybe even service pack 3 by now. NOTE: When installing your service packs, if the PC asks you to reboot, do it then, don't wait.

Insert your Motherboard CD/DVD, and install all the drivers, and if you want, the optional software. If it asks you to restart the computer, go ahead.

If you have a video card, or an audio card, Use the installation CD/DVD's that came with those. Also install any printer, scanner, or all-in-one software CD/DVD's.

Now would be a good time to make a "my computer" icon on the desktop. Press start, go over to the "my computer" tab, and right click on it. Select "show on desktop". Now go to your desktop, and right click on the "my computer" icon we just made. Select "manage", and go down to "device manager", and left click on that. The device manager should show all of your hardware.

Satisfying the device manager
Look at all the hardware lines in the device manager, and if any of them have a yellow question mark next to them, we need to install drivers for these devices. Remove your motherboard CD, and replace it with your driver disc that you made in the
SOFTWARE INSTALLATION- MAKING A DRIVER DISC section. If your PC is still asking for drivers, click "have disc" then point to your D: drive, and follow the directions on your PC. If you do not have any yellow question marks, then we're all set. For some of these drivers, you will need the driver disk you made. Your PC

may want to restart, if so do it. Check your device manager again. At this point we should have no question marks left.

NOTE: If you have a red X, then the device is bad, or it needs installation.

<u>Ways to install and re-install drivers</u> – Sometimes windows fails to automatically install drivers.

1. Self installing. -- This is the easiest driver to install, you just click on it. Video and Audio drivers are usually this type.
2. Need to go thru the add hardware wizard. -- Go to the control panel, and select add hardware. The add hardware wizard will walk you through it.
3. Uninstall then re-install. -- Sometimes you need to remove a driver before you can add another one. To remove a driver, go to the device manager, and remove it from there. Right click on the driver, and select un-install.

NOTE: When you un-zip drivers, it will ask you where you want to extract the files to. Make a folder called Unzipped on your destop, or someplace that you can remember like your C: drive. DO THIS BEFORE YOU UNZIP A FILE. Then extract the files to this location so you will know where the files will be when you are done unzipping.

4. Zipped drivers. – Simply click on the driver, and un-zip it.

NOTE: Don't forget to install the processor driver you placed on your CD/DVD that you made. If your PC is running smoothly, then do not update your bios, or drivers.

NOTE: ONLY update ONE driver at a time, if need be.

NOTE: If your PC is unstable, or keeps crashing, update your video driver first. Use your disk that you made. Check to make sure everything is okay, run it for a while, and if you still have problems then update your audio driver second. If one particular piece of software keeps crashing, or does not work well, then look for a patch for it on the internet, or the game website. Gamespot is also another good place to get game patches.

NOTE: A FEW WORDS ON DRIVERS. The saying "newer is better", is not necessarily true of software, or hardware drivers, or bios. IF ALL IS WELL, THEN DON'T TOUCH. If after updating something, and your PC doesn't seem to run as stable as it use to, then either rollback, or remove the driver, or piece of software that you just installed. If you added software, then check to see if there is an update for it on the web.

NOTE: Notice the big note. ← If all fails, and the device manager is satisfied, and you have an unstable system, then update the bios. Read your mobo book on how to do this. DO NOT WAIVER FROM THE PROCEDURE because you may lock up your bios, and need to have it serviced.

Chapter 12

INSTALLING A SECOND VIDEO CARD

Power down, and turn off the power to everything. Refer to the <u>Install video card</u> section, chapter # 6. Install your second video card, screw it in, then power up and setup the second video card with drivers. Check with the device manager to be sure they both got setup correctly. To see the device manager, click on start, right click on "my computer", click on "manage", then click on the "device Manager".

NOTE: Don't forget the little SLI bridge connector between the video cards, IF you are going to use SLI mode. Also the video cards have to be the same model. You also have to set the PC to SLI mode. The PC should detect the second identical video card, and ask you if you want to set up SLI mode. IF the PC does NOT ask you this, you will have to set up the SLI mode manually. Go into the video card software control panel to do this. Click on start, click on "all programs", click on the video card graphics processor manufacturer, (either Nvidia®, or Ati®), (look on your video card box if necessary), and select the video cards control panel. Choose SLI mode, and turn on SLI mode.

Chapter 13

Windows® XP re-installation

Use this procedure if you have to re-install your operating system for some reason.

NOTE: WRITING ZEROES TO A DRIVE IS USED TO COMPLETELY ERASE A
DRIVE. Should you desire to do this ?

NOTE: You should write zeroes to your hard drive before reinstalling XP® You can use
the utilities that came with your Western Digital® hard drive. This will ensure a clean
installation. This will take several hours, and after it starts, it does not need any
interaction.

Insert the windows® installation disc, and power up while holding the space bar.
Some files will load automatically, just wait.
At the first prompt press enter.
Press F8.
If a previous windows® installation is present press ESCAPE, otherwise press ENTER.
Make sure C: is highlighted, then press ENTER.
Press C.
Choose format the partition using the NTFS file system (quick).
Press ENTER.
Press F.
Setup should check the disc, and then start to copy files.
Now go to section <u>Windows® Basic Installation</u> XP step # 7. (Chapter 11)

Chapter 14

TESTING

This would be a good time to test your voltages, and temperature again, as your system is up to temp by now. Restart your PC, and hold the delete key again to get into your bios screen. Go thru the voltage and temperature readings again, and check them to see if they are okay.

Fan settings

If you can adjust your case fan speeds,(if they have a speed switch), lower the speed from high to medium. In 15 minutes, check (thru your bios) to see if the temperatures go up dramatically. If so, then leave the fan speed on high. If the temperature stays pretty much the same, then leave it at medium. This will decrease overall system noise, and dust.

ANTIVIRUS

Installing a good antivirus is essential for safely surfing the web these days. Scanning your system for viruses is also a good idea. See WEEKLY MAINTENANCE. The only beef I have with Norton is that they consider a tracking cookie to be a "low level" threat, I don't. This after all is telling someone everywhere you go on the internet.

Chapter 15

FINISHING UP

<u>Direct X®</u>

Direct X® is needed for most games, and multimedia applications. Applications, or games that need a Direct X® update usually come with a version of direct X®.
NOTE: Even if you have already installed a version of Direct X, if a program asks you to install it, click YES. The reason for this is: There are different Direct X® instructions with different programs, even though the version may be the same.

<u>DONE</u>:

<u>CONGRATULATIONS</u>, you have finished building, and installing the software on your PC, ENJOY !!!

Chapter 16

GAMES

<u>Installing in Separate Folders</u> –

Games like to make a mess of your C: drive. If you need to remove a game, sometimes stuff stays behind on your hard disc. What I do is make a separate folder for each game on my system. Then I install the games in their own separate folders. I have a folder called Games on the root of my hard drive. (C:\Games), and I make the game folders in this one. This way if the game does not un-install itself correctly, I can remove the separate game folder, and it is completely gone. To make the game folder, click on start, my computer, double click on C:, under file click new, click on folder, then call it Games, and press enter.

<u>To Calibrate Controllers</u> –

Sometimes windows® loses its game controller calibration. You will notice this if you are playing a game, and the controller drifts, or as in a flying game, the aircraft does not fly straight. Microsoft flight simulator X sometimes also loses it's calibration. To calibrate your controller, click start, go to the control panel, and double click game controllers. You will see your controller listed, select it, click properties, click settings, and select calibrate. Follow the instructions in the calibration wizard. Click apply, click okay.

Chapter 17

MAINTENANCE

WEEKLY MAINTENANCE
Antivirus signature update
Scan for viruses

MONTHLY MAINTENANCE

Defragmenting the Hard Drive
Run Defrag program. If the system starts to slow down, do it more often than once a month.

Defragmenting the Registry
"Aus logics registry defrag" is a neat tool. If you experience system slowdown and defrag doesn't speed it up. You either have a virus and/or your registry is fragmented. This happens if you are constantly adding and deleting programs and files. You can find it on the web for a fair price.

6 MONTH MAINTENANCE

CLEANING

Remove the trek ball ball, and clean rollers vigorously with a Q-tip. Make sure the rollers actually roll !!! Roll them with your thumb to make sure there not stuck with gunk ! Now replace the trek ball ball. Ahhhh, doesn't that roll easier !

Take out and wash air filters (if you have them).

Clean fan blades, and air intakes.

Clean Scanner glass.

Chapter 18

BACKING UP

You should always keep a backup of things that you do not want to lose. If you have DVD burning software, you can store 4.7 Gigabytes of pictures, songs, and documents, etc… on a recordable DVD. If you have Dual layer capability on your DVD drive, then you can store almost twice this amount of data. A few words of advice, if you don't want to lose it then store it ! What would you do if your hard drive failed right NOW !!! Would you have what you NEED ?

Chapter 19

GENERAL TROUBLESHOOTING

Is it on ? Laugh all you want, this is the first rule of troubleshooting.
Is it plugged in all the way ? is another.

AUDIO

Speakers do not work -

Are they in the right jack on your sound card. Are the speakers on ? Are the speakers plugged in ? Go to the control panel icon (Sounds and Audio devices). See if the volume has been turned off, or lowered. Check and see if your mixer is set up properly with the check boxes on the Wave, and midi checked. Check the device manager for yellow question marks, or red X's. If you have one of these, then reinstall the sound driver. If you still don't have audio, then try another set of speakers. If you still don't have audio, turn off your system, turn off the power, and remove your sound card and re-install it. If you still don't have audio, your sound card is probably faulty.

Surround Sound-

Most new games have surround sound. If you have surround sound, you must go into the control panel, and setup the speakers to surround sound. If you have a sound card, you must setup the speakers in both the windows® audio icon, and the sound card's icon in the control panel. You must also check the box marked "synchronize with windows® control panel", under the speakers tab, or you won't have surround sound ! You also have to set the speaker arrangement to 5.1, or 7.1/8 speakers if you have side speakers along with the center speaker and subwoofer.

Video

If you have no video, is your monitor on and the brightness and contrast turned up ? Try another monitor in it's place. If you still have no video, turn everything off, then remove your video card and put it back in. NOTE: see figure # 4 for video card removal. It may not have been fully seated, and screwed into the socket. If you still don't have video, then maybe your video card is faulty ?

Mouse or Trek ball

Make sure it is plugged in all the way. Check the device manager, for your mouse or trek ball, to make sure it is listed. If you installed mouse or trek ball software, you can try removing the software driver, and re-install it.

Printer

Are there any blinking lights on your printer ? Is it out of paper, or ink ? Paper jam ? Check the device manager to make sure your printer is listed. You can also clear a stuck document by clicking on the start button, then double click on the control panel. Double click on the "printers and faxes" icon. Double click on your printer icon, then click on document, and then click cancel. Try unplugging the USB cable, wait a few seconds then plug it back in while your printer is on. You can also try un-installing, then re-installing the printer driver, maybe it got corrupted.

USB Devices

Windows® sometimes only recognizes a USB device if, the device is turned on before turning on your PC. Sometimes you need to plug in your device after windows is booted up. It depends on the device, and whether the USB device requires power from an external source. Check the device manager to make sure your USB device is listed after you plug it in. After you plug in a USB device, windows should acknowledge the device by making a sound.

Updating Drivers

If your peripheral doesn't work like it's supposed to, then use your driver disc to update the driver. See Ways to install and re-install drivers Chapter 11.

Chapter 20

FREQUENTLY ASKED QUESTIONS

<u>Should I leave my PC on all the time ?</u> – The way I look at is if you're not using it, it is wasting electricity, and shortening the life of your PC. Heat and dust are PC's worst enemies. If you are a business, then of course, you need it on all the time.

<u>How do I find out how full my hard drive is</u> ? Press start, left click on my computer, right click on local disc C:, and choose properties.

<u>Order of system power up</u> ? Power strip, Monitor, Printer, Speakers, PC.

<u>Order of power down</u> ? PC, Speakers, Printer, Monitor, Power strip.

<u>Over clocking ?</u>
   Years ago you could clock an Intel® Celeron® at almost twice it's speed. Now processor testing falls into very strict categories. Today 3% is pushing it. I don't recommend over clocking because if your system hiccups, murphy's law will come into effect, and destroy your most critical part of your operating system. Now you will have to re-install windows, and all your programs all over again. Unless of course you have a disk image on another hard drive as a backup. (RECOMMENDED) System restore to the rescue, forget it !!! This is why I recommend 2 hard disc drives, one for backup.

Chapter 21

Products and Companies: In order of appearance

MICROSOFT XP®
MICROSOFT WINDOWS XP64
MICROSOFT VISTA®
NVIDIA®
CORSAIR®
AMD ATHLON 6400 DUAL CORE PROCESSOR®
ASUS
WESTERN DIGITAL HARD DRIVES
CREATIVE SOUNDBLASTER®
CREATIVE SOUNDBLASTER® X-FI® EXTREME GAMER SOUND CARD
CREATIVE SOUNDBLASTER® X-FI® EXTREME MUSIC SOUND CARD
EVGA / NVIDIA GEFORCE 9600GT® VIDEO CARD
EVGA/ NVIDIA GEFORCE 9800GTX+® VIDEO CARD
SONY DVD DRIVE
AMD
AMD AM2
AMD AM2+
AMD ATHLON 64 BIT DUAL core®
ASUS CROSSHAIR®
ANTEC FANS
MEMOREX DVD DRIVE
INTEL
ATI
ANTEC CASES
LOGITECH MODEL X-540 SPEAKERS
INVAR® SHADOW MASK
PHILIPS® MODEL 201B4 TUBE MONITOR
MICROSOFT NATURAL® KEYBOARD
LOGITECH MARBLE MOUSE
EPSON STYLUS PHOTO® MODEL R320 PRINTER
EPSON EXPRESSION® 1600 SCANNER
YAMAHA SPEAKERS
ALTEC LANSING SPEAKERS
RADIO SHACK
MICROSOFT WINDOWS®
THERMALTAKE HEATSINK FANS
NORTON

MICROSOFT DIRECT X®
AUS LOGICS REGISTRY DEFRAGMENTATION PROGRAM
Q-TIP®
INTEL CELERON®

NOTE: I do not own stock in any of the above companies, nor did I accept payment or merchandise for advertising any company name or product.

About the Author

Mark lives in Massachusetts. He has 28 years of PC computer experience, with an Associates Degree in applied Science, with an Electronics major. He is co-inventor of 7 US Patents, and Co-author of two published papers. Mark enjoys, PC simulators (especially flying), 3d shooter games, and working with electronics.

Motherboard   Make:
                   Model:
                   Serial #
                   Revision #

Processor   Make:
                   Model:
                   Serial #
                   Speed in Gigahertz:

Video Card   Make:
                   Model:
                   Serial #
                   Video memory in Megabytes:

Hard Drive   Make:
                   Model:
                   Serial #
                   Size in Giga or Terabytes:

CD/DVD   Make:
                   Model:
                   Serial #

Memory   Make:
                   Model:
                   Giga Bytes total:

Monitor   Make:
                   Model:
                   Serial #

Power Supply   Make:
                   Model:
                   Serial #
                   Wattage:

Printer   Make:
                   Model:
                   Serial #
                   Ink Cartridge #'s

NOTES

NOTES